# *Argh!*

## *to*

# AHA!

## Math Made Easy

*John Diamondidis*

**CITIOFBOOKS, INC.**
3736 Eubank NE Suite A1
Albuquerque, NM 87111-3579
*www.citiofbooks.com*
Hotline:      1 (877) 389-2759
Fax:          1 (505) 930-7244

Ordering Information:
Quantity sales. Special discounts are available on quantity purchases by corporations, associations, and others. For details, contact the publisher at the address above.

Printed in the United States of America.
ISBN-13:      Paperback      979-8-89391-763-5
              eBook          979-8-89391-764-2

Library of Congress Control Number: 2024916690

# Argh!
## to
# AHA!
## Math Made Easy

*John Diamondidis*

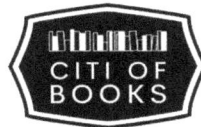

CITI OF
BOOKS

# CONTENTS

Introduction

1. Addition Made Simple

2. Subtraction Made Simple

3. Staying with Subtraction Problems

4. Word Problems Made Simple

5. Multiplication Made Simple

6. Staying with Multiplication

7. Division Made Simple

8. Solving Division Problems

9. More Division!

10. Fractions Made Simple

11. More on Decimals

12. Percentages Made Simple

In Conclusion: From One Math Struggler to Another...

# INTRODUCTION

Hi, I'm John Diamondidis.

Most of you don't know me. So, allow me to give you a little background on me.

I'm 59 years old. When I was in Elementary School I was labeled as HYPER ACTIVE. Today's time, I would have been labeled ADD and prescribed medicine to get through classes.

All my teachers used to tell my parents how nice of a kid I was. They said all the students liked me. They also said I was not grasping the information fast enough, and I was going to be held back if I didn't grasp MATH!

After my parents told me this, I really paid attention, especially during math classes. It wasn't working. So I took a little of this and a little of that and created my own system for math—adding, subtracting, multiplying, and dividing. **I found that my methods worked for me!**

I went from being nearly held back to an **All–Star** in math. Math is kind of cool anyway. There are no trick answers. Math is math. 2 + 2 = 4...always! No other answer is correct! So I figured out my own methods of coming up with the correct answers, and I was always able to show and explain my work!

I was asked by a friend to explain my method of math in written format, so others could benefit from what I found to be very simple.

*I practiced counting to 100 by ones, twos, fives, and tens! This knowledge helped me figure out a simpler way to do math problems. I created several methods. One of which I call "Go with what you know!"*

**The most important thing about mathematics to know and understand is that our whole number system is BASED on 10. When in doubt, use your fingers and toes.**

| Ones (1s) = | Tens (10s) = | Hundreds (100s) = | Thousands (1,000s) = |
|---|---|---|---|
| 0 – 9 | 10 ones | 10 tens | 10 hundreds |

Do you see what I'm talking about? Practice counting to 100 by ones and tens, and you will understand.

> See how many ones it takes to get to 10.
> Count how many tens it takes to get to 100.

Also, something to remember is that **5 is half of 10**! Always. No exceptions. 5 is half of 10. Why is this last fact so important? Because it holds true for every 10!

> Half of 10 is 5.
> Half of 100 is 50.
> Half of 1,000 is 500!
> This is the fact of math!

**So, it is highly important that you can count to 100 by ones, fives, and tens!**

Remember, if it helps you to keep things straight by writing your numbers bigger on the work paper, then write them bigger. It always helped me.

4 or **4**

It doesn't matter how big you write the numbers, as long as you work the work and get the correct answer. Every pencil has an eraser. If you get confused or don't get the correct answer the first time...Use your eraser.

**Trust yourself!**

**Okay, let's get started!**

# 1

# ADDITION MADE SIMPLE

Always remember, zero added to any number is that number. When you add zero to any number, you are adding NOTHING to that number.

0 + 1 = 1    0 + 2 = 2    0 + 3 = 3    0 + 4 = 4    0 + 5 = 5
0 + 6 = 6    0 + 7 = 7    0 + 8 = 8    0 + 9 = 9    0 + 10 = 10

Know that no matter how you write an addition problem, if the same numbers are used, the answer is the same.

1 + 0 = 1    2 + 0 = 2    3 + 0 = 3    4 + 0 = 4    5 + 0 = 5
6 + 0 = 6    7 + 0 = 7    8 + 0 = 8    9 + 0 = 9    10 + 0 = 10

The same goes for when the number are stacked:

| 1 | 0 | 0 | 4 | 7 | 0 |
|---|---|---|---|---|---|
| +0 | +1 | +4 | +0 | +0 | +7 |
| 1 | 1 | 4 | 4 | 7 | 7 |

Now, ADDITION is just counting. This is pretty cool. If you can count to 10, you can answer any ADDITION problem. Remember, you have 10 fingers and 10 toes. If you need them to solve a math problem, use them! Addition is easy if you SEE the numbers in your head.

ADDITION problems are all how you see the math problem. I found it easier to change all addition problems **to stacked problem:**

| YOU SEE | I SEE | YOU SEE | I SEE | |
|---|---|---|---|---|
| 8 + 2 = | 8<br>+2 | 8 + 20 + 113 = | 113<br>+20<br>+8 | **When writing the ADDITION problem, I like it better with the biggest number on top.** |

REMEMBER, with addition problems it doesn't matter how it is written. 2+2=4... always.

So when solving the addition problem 6+20+113, **Remember:** no matter how you write the addition problem, the answer will be the same. My method is to break the big addition problem into three easier problems. Also remember that 0 (zero) plus any number is that number!

| YOU SEE | I SEE | HUNDREDS | TENS | ONES |
|---|---|---|---|---|
| 6 + 20 + 113 = | 113<br>+20<br>+6 | 100 | 10 | 3 |
| | | +000 | +20 | +0 |
| | | +000 | +0 | +6 |
| | | 100 | 30 | 9 |
| | | +30 | | |
| Add the three numbers **together:** | | +9 | | |
| | | 139 | | |

Using this method, I found it easier and faster because I am adding a lot of numbers to zero. This keeps it very simple by breaking a big number into smaller numbers and adding.

Sooner or later, you'll be able to see the numbers in your head (mind) and do the math with less rewriting of the addition problems. **Remember,** keep your columns straight.

# 2

# SUBTRACTION MADE SIMPLE

Subtraction is as simple as counting backwards from 10. Again, our number system is based on tens. Practice counting backwards from 10.

10    9    8    7    6    5    4    3    2    1    0

Subtraction problems are just "take away" problems. It just depends on how you read them.

**YOU SEE**      **I SEE**
**9 – 7 =**       **9 take away 7 =**

I start at 9 and count backwards. Here's the thing if you're gonna count backwards:

When you say the word "nine," make sure you have both your hands closed when you say 9. Then start raising your fingers until you have 7 fingers up.

**An easier way to look at subtraction problems:**
Keep in mind, you still start by subtracting the ones column first, then the tens column. Then add the two numbers together to get the correct answer.

| YOU SEE | I SEE | | |
|---|---|---|---|
| | TENS | ONES | |
| 27 − 15 = | 20 | 7 | |
| | −10 | −5 | |
| | 10 | 2 | |
| | +2 | | Once you have the answer for both columns, ADD the two columns together to get the answer! |
| | 12 | | |

If this simple method does not work for you, I'm about to share another way. The way I'm about to share with you I call the "GO WITH WHAT YOU KNOW" method. Parents, you should/can work with your kids on this. This method takes a lot of practice counting. By practicing, I became very good at counting by ones, twos, fives, and tens. Practice counting to 100!

**KNOW THIS: A subtraction problem is an addition problem written differently.**

| YOU SEE | I SEE | | |
|---|---|---|---|
| 37 − 19 = | 37 | | I also see a chance to "go with what I know!" **19 + what number = 37?** I know that 19+1=20 (I know this because I counted by ones). So, I write down 1. |
| | −19 | | |
| | 19 + | 1 | I know that 20+10=30 because I counted by tens. I know that 30+7=37 because I counted by ones. Add the numbers up to get the answer to 19 + what number = 37. |
| | | +10 | |
| | | +7 | The answer to 19 + what number = 37 is 18! |
| | | 18 | |

Knowing that 19 + 18 = 37, **we also know that 37 − 19 = 18.**

We also know that 37 − 18 = 19.
We also know that 18 + 19 = 37!

# 3

# STAYING WITH SUBTRACTION PROBLEMS

If you prefer the stacking method instead of "going with what you know," I want to show you the following, so you know what to do.

**Again, start solving the problem by doing the ONEs column first.**

| YOU SEE | I SEE | | |
|---|---|---|---|
| | **TENS** | **ONES** | Just by looking at this problem, you know you cannot take 9 away from 1. Wathcha you gonna do, Scooby–Doo? I'll tell you: you take 10 away from the TENS column and add it to the ONES column. Then, rewrite your problem. This is why each pencil has an eraser! |
| 51 – 19 = | 50 | 1 | |
| | −10 | −9 | |
| | **TENS** | **ONES** | **Now do the math!** |
| | 40 | 11 | |
| | −10 | −9 | |
| | 30 | 2 | (Now, Add the two columns together) |
| | +2 | | |
| | 32 | | **Remember, add the totals of your columns to find the answer.** |

**Another way to tackle 51–19= is to "go with what you know!"**

**51–19=?** Or look at it this way: 19 + what number = 51? (We turn the subtraction into an addition).

Let's solve: We know that 19+1=20. How do we know this? We started counting from 19 to 20.

Then count by tens to go from 20 to 50, and you get three tens (30).

| | 20 to 50 | 30, 40, 50 | |
|---|---|---|---|
| 51 – 19 = | | 10 + 10 + 10 | Use your fingers to keep track if you need to. |
| | 1 | | So you get 30...write it down. |
| | +30 | | We're almost there...count by ones to get to |
| | +1 | | 51, write it down! |
| 51 – 19 = 32 | | | **Now just add the numbers to get the answer.** |
| | 32 | | |

**NOW WE KNOW** the answer to 19 + what number = 51. We know the answer is 32! So, we also know that 19+32=51! And we know that 32+19=51!

You may be asking yourself, "What is the answer to 51–19?"

Well, let's take a look. Remember that a subtraction problem is just an addition problem written differently.

So, by solving 19 + what number = 51, **we also solved the problem of 51–19 = ?**

   **51 – 19 = 32!**

By solving 19 + what number = 51, we now know that:

$$51 - 19 = 32$$
$$51 - 32 = 19$$
$$19 + 32 = 51$$
$$32 + 19 = 51$$

Look how many math problems we just solved. It all depends on how you look at the problems!

**GOOD JOB!!!**

# 4

# WORD PROBLEMS MADE SIMPLE

*Johnny has 10 apples, 4 oranges, and 6 bananas. Sally takes 4 apples from Johnny. Timmy takes 2 bananas from Johnny. How many pieces of fruit does Johnny have?*

Okay, first look at the question. How many pieces of fruit does Johnny have? So, now I read the whole thing. As I am reading, I make columns. Each time the question talks about Johnny and fruit, I write it down!

| | Apples | Oranges | Bananas | |
|---|---|---|---|---|
| **Johnny** | 10 | 4 | 6 | |
| **Sally** | −4 | −0 | −0 | I keep reading. Sally takes 4 apples from Johnny. |
| **Timmy** | −0 | −0 | −2 | Timmy takes 2 Bananas from Johnny. |
| | 6 | 4 | 4 | **I only care about what people take or give to Johnny. Because the question is about Johnny!** Now, do the math in each column. Keep in mind, if there is no entry you can put a zero (0). |
| | 6 | | | |
| | +4 | | | |
| | +4 | | | Now is all we have to do is to add the totals of |
| | 14 | | | the columns! |
| | **14 pieces of fruit!** | | | **Now, label your answer.** |

My biggest challenge with word problems is that I am a slow reader. This method sped it up and helped me keep organized. When labeling your columns, you can use abbreviations. For example: Instead of writing APPLE, you can label it "A," ORANGE with an "O," BANANA with a "B"!

Word problems can get bogged down with no real meaning. But let's simplify all. You see the following word problem:

*Johnny has 4 apples, 6 oranges, and 12 bananas. Sally takes 4 apples from Johnny and gives him 6 bananas. Jake takes 2 oranges from Johnny and 3 apples from Sally. How many bananas does Johnny have left?*

**Let's solve this!**

Read the end of the question first:

*How many bananas does Johnny have left?*

Okay, the question is about bananas and Johnny! SO, THAT'S ALL YOU HAVE TO BE CONCERNED ABOUT...BANANAS!

| | Bananas | |
|---|---:|---|
| Johnny | 12 | Johnny starts with 12 bananas. |
| Sally | +6 | Sally gives Johnny 6 more bananas. |
| | 18 | |

What about Jake? What Jake does doesn't matter because he didn't do anything with bananas! The question is: "How many bananas does Johnny have left?"

**The answer is: Johnny has a total of 18 bananas.**

**EASY-PEASY!**

# 5

# MULTIPLICATION MADE SIMPLE

Multiplication problems are just "quicker" ways to do addition problems. While multiplying, there are ultimate rules that you must remember:

• Any number multiplied by zero (0) equals zero! It can be 1,340,893 x 0; the answer is 0!

• Any number multiplied by 1 is that number:

456 x 1 = 456

85 x 1 = 85

**12 x 1 = 12**

Okay, let's make multiplication EASY. Remember that multiplication is a shortened way of writing an addition problem...

**You may be asking yourself, "What the heck is he talking about!?"**

| YOU SEE | I SEE | YOU SEE | I SEE | I ALSO SEE |
|---|---|---|---|---|
| 6 x 4 = | 6 | 13 x 13 = | 13 | |
| | +6 | | +13 | 39 |
| | +6 | | +13 | |
| | +6 | | +13 | |
| | 24 | | +13 | +39 |
| | | | +13 | |
| YOU SEE | I SEE | | +13 | |
| 5 x 8 = | 8 | | +13 | +39 |
| (or 8 x 5 =) | +8 | | +13 | |
| | +8 | | +13 | |
| | +8 | | +13 | +39 |
| | +8 | | +13 | |
| | 40 | | +13 | +13 |
| | | | 169 | 169 |

14

Okay, now you, too, can see multiplication problems as addition problems. As you get more practice, you will be able to do multiplication problems as they are written—without changing them into addition problems.

Again, just like with addition problems, I like to stack the multiplication problems. Also, the numbers can be placed on either side of the "X." Example:

6 x 8 = 48          8 x 6 = 48

4 x 2 = 8           2 x 4 = 8

When stacking multiplication, I like to put the biggest number on top!

| YOU SEE | I SEE | YOU SEE | I SEE | I ALSO SEE | |
|---|---|---|---|---|---|
| | | | | TENS | ONES |
| 6 x 8 = | 8<br>x 6<br>48 | 8 x 12 | 12<br>x 8<br>96 | 10<br>x 8<br>80<br>+16<br>96 | 2<br>x 8<br>16 |

# 6

# STAYING WITH MULTIPLICATION

There is another rule besides the 0 rule and the 1 rule. This will help things go a lot faster.

**THE RULE OF 10!**
Any whole number multiplied by 10 is that number with a zero added to the end!

1 x 10 = 10        2 x 10 = 20        3 x 10 = 30        4 x 10 = 40        5 x 10 = 50
6 x 10 = 60        7 x 10 = 70        8 x 10 = 80        9 x 10 = 90        10 x 10 = 100
32 x 10 = 320      14 x 10 = 140      6,253 x 10 = 62,530!

**The Rule of 100!**
100 is like 10 but it has two zeros instead of one, so we know that any number multiplied by 100 is that number with two zeros at the end!

6  x 100 = 600
12 x 100 = **1,2**00
9  x 100 = **9**00
10 x 100 = **1,0**00
46 x 100 = 4,600
**...and so on.**

The same rule applies when multiplying by 1,000; 10,000; 100,000; and 1,000,000.

However many zeros there are, add them to the end of the number, and that's the answer.

Okay. This next concept is one of my favorites. It makes multiplying very quick and simple. It is a little trickier, but you'll see the benefits.

Quick Cheat: **My Dear Aunt Sally**.

When math problems become more complex just remember My Dear Aunt Sally.

EXAMPLES:

6 + 9 x 2 − 4 ÷ 2 =

First, do the multiplication (My):   9 x 2 = 18
Then do the division (Dear):        4 ÷ 2 = 2
Now, rewrite the problem:    6 + 18 − 2 =
Now add (Aunt):        6 + 18 = 24
Rewrite the problem:   24 − 2 =
Now subtract (Sally):   24 − 2 = **22**

## GO WITH WHAT YOU KNOW:

**YOU SEE     I SEE**
**13 x 9 =     13 x 10 - 13**

Okay, let me explain what I see:

13 x 9 = is the same as 13 x 10 − 13!

First, you multiply 13 x 10. We do this because we KNOW that 13 x 10 = 130. However, the problem is 13 x 9. Basically, 13 x 9 means how much does nine thirteens add up to? Well, we know that 10 thirteens is 130. Nine is one less than 10, so we figure 10 thirteens is 130.

To find nine thirteens, we now subtract one thirteen from 10 thirteens to get the answer. Remembering My Dear Aunt Sally, the math problem becomes:

**13 x 10 − 13 = 117**

Now, I'll show you what I'm talking about, step by step!

| YOU SEE | I SEE | | | | | |
|---|---|---|---|---|---|---|
| 13 x 9 = 117 | 13 | | | | | |
| | x 10 | | | | | |
| | 130 | We know this because of THE RULE OF 10. | | | | |
| | −13 | But the question is 13 x 9 = ? | | | | |
| | 117 | So 9 is one less than 10. So, we know that we have to subtract one 13 from 13 x 10 to get the answer to 13 x 9. | | | | |
| | | | | | | |
| 12 x 8 = | 12 | and/or | 12 | and/or | 10 | 2 |
| | x 10 | | x 10 | | x 8 | x 8 |
| | 120 | | 120 | | 80 | 16 |
| | −12 | | −24 | | +16 | |
| | 108 | | 96 | | 96 | |
| | −12 | | | | | |
| | 96 | | | | | |

## RULE OF 5!

When multiplying any number by 5 or any number that ends in a 5, the answer will end in a 0 or a 5. Any **even** number multiplied by a 5 will end in a 0. Any **odd** number multiplied by 5 will end in a 5!

7 x 5 = 35     4 x 5 = 20     13 x 5 = 65     20 x 5 = 100     15 x 5 = 65     12 x 5 = 60

## ODD OR EVEN RULE OF MULTIPLICATION:

EVEN Number x EVEN Number = EVEN Number:
2 x 2 = 4     12 x 8 = 96     14 x 4 = 56

ODD Number x EVEN Number = EVEN Number:
3 x 2 = 6     9 x 4 = 36     11 x 8 = 88

ODD Number x ODD Number = ODD Number:
5 x 7 = 35     3 x 15 = 45     9 x 11 = 99

# 7

# DIVISION MADE SIMPLE

Division problems are just multiplication problems written differently, just like subtraction problems are addition problems written differently!

**Rules to know while dividing:**
- Zero (0) divided by any number is zero.
- Any number divided by 0 is undetermined!

**EXAMPLES:**

$0 \div 4 = 0$          $0 \div 17 = 0$

$24 \div 0 =$ **undetermined**      $56 \div 0 =$ **undetermined**

| Undetermined | Undetermined | 0 | 0 |
|---|---|---|---|
| $0\overline{)7}$ | $0\overline{)45}$ | $45\overline{)0}$ | $56\overline{)0}$ |

When reading a division problem, you have to know how to read it in order to solve it! Once you know how to read it, you can DO IT! When you see this symbol $\div$ in a math problem, you must read it as **"DIVIDED BY."**

**Example:**
$12 \div 6 =$

You read by saying, "**12 divided by 6**." If you're in class, or taking a test, read in your 'whisper voice.'

**YOU READ** the example as follows: "12 divided by 6 equals…"
**I READ:** "6 goes into 12 how many times?"
**OR:** "6 times what number = 12?"
I always write division like this:

$$6\overline{)12}$$

I like doing all my division work in the "goes into" format, which is also known as LONG DIVISION ($\overline{)}$ ).

**YOU SEE**          **I SEE**

$45 \div 5 =$          $5\overline{)45}$

45 divided by 5 =          5 "goes into" 45 how many times?

Both division problems above ask the same question but are written differently! I prefer working division problems written as in the "I SEE."

On occasion, you will see a "/" used for division instead of the ÷ or the $\overline{)}$ sign. The / is usually used when working with fraction and so forth. (More on fractions later.)

# 8

# SOLVING DIVISION PROBLEMS

Okay, here we go! We're going to start easy and then get a little more difficult. No biggie; it's a matter of getting to the correct answer!

| YOU SEE | I SEE | |
|---|---|---|
| 75 ÷ 15 = | 15 )‾75 | When we look at this division problem, I FIRST see that 15 ends with a 5, and the number we're dividing it into is 75, which also ends in a 5. Therefore, I know the answer will end in an odd number. 5 is an odd number.  Odd divided by odd always equals odd! |

**How many times does 15 go into 75?**

I first start out by asking myself, **"What is 15 x 2?"** I know it is the same as asking **15 + 15, which is 30**. Once we know **15 x 2 = 30**, we know that **15 x 4 = double 30, which is 60**. Now, **how many more fifteens can we fit into 75?**

| 15 x 2 = | 30 | |
|---|---|---|
| | +30 | |
| 15 x 4 = | 60 | Can 75 divided by 15 be 4? No, it can't, because 15 x 4 = 60, and 75 is our goal. |
| | +30 | How many fifteens can we stuff into 75? |
| 15 x 6 = | 90 | Can 75 ÷ 15 be 6? No, it cannot be, because 15 x 6 = 90. So let's try 75 ÷ 5. |
| | −15 | Instead of rewriting everything, and since our last try went over 75 when we did 15 x 6, let's try subtracting one 15 from 15 x 6 to see |
| 15 x 5 = | 75 | what 15 x 5 equals. |

**BOOM! We know that 75 ÷ 15 = 5!** We also solved several other problems:

$$75 ÷ 15 = 5$$
$$75 ÷ 5 = 15$$
$$5 \times 15 = 75$$
$$15 \times 5 = 75$$

**GOOD GOING!** It is all a matter of how you see the math problems. It is easier for me to add and multiply than it is to subtract and divide! It may be for you too!

**More Division Practice**

| YOU SEE | I SEE | I ALSO SEE |
|---|---|---|
| 108 ÷ 12 = | 12 x what number = 108 | 12 ⟌ 108 |

I solve the same way as we did before. I solve it by using the tried-and-true way...FOR ME!

**I SEE**

12 x what number = 108?

(Again, the math problem we are about to solve is: 108 ÷ 12 = ?)

Let's start solving this sucka! What is 12 x 10? Remember the RULE OF TEN in the multiplication part of this book. When multiplying any whole number by 10, just add a zero at the end of the number! We'll solve this math problem without doing much math. **GO WITH WHAT YOU KNOW!**

| | | |
|---|---:|---|
| 12 x 10 = | 120 | 120 is more than 108, so we know that **the answer is not 10, but we're close! Subtract one twelve from, 120 and what do we have?** |
| 12 x 10 = | 120 | |
| | − 12 | |
| 12 x 9 = | 108 | So, **if 12 x 9 = 108, we know that 108 ÷ 12 = 9!** We just took a giant leap in **"going with what you know"** to make math easier! And by solving the math problem 108 ÷ 12 = 9, we also know that 108 ÷ 9 = 12! We also know that 12 x 9 = 108 and 9 x 12 = 108! |

## GOOD JOB! VERY GOOD JOB!

# 9

# MORE DIVISION!

Division problems will sometimes have a remainder. A remainder is nothing more than leftover numbers. Let's take a look at the "remainder" situation!

| YOU SEE | I SEE | I ALSO SEE |
|---|---|---|
| $25 \div 7 =$ | $7\overline{)25}$ | 7 x what number = 25 |

Let's Figure it Out!

| | | |
|---|---|---|
| 7 x 1 = 7 | 7 | So now we know 7 x 3 = 21 and 7 x 4 = 28. |
| 7 x 2 = 14 | +7 | We now know that 7 does not go into 25 evenly. |
| 7 x 3 = 21 | +7 | No big deal...here's what we do! |
| 7 x 4 = 28 | +7 | |

```
    *       3r4
  7      ⌐25
 x 3     − 21
  21        4
```

We know we cannot divide 4 by 7. So the 4 becomes the remainder.

**And the answer to  25 ÷ 7 = 3r4!**

EXAMPLE:

| YOU SEE | I SEE | |
|---|---|---|
| | | 4r8 |
| 124 ÷ 29 = | 29 $\overline{)124}$ | |
| | +29    -116 | |
| 29 x 2 = | 58         8 | |
| | +29 | |
| 29 x 3 = | 87 | |
| | +29 | |
| 29 x 4 = | 116 | |

Okay, this is a difficult division problem. Let's solve this one together! There's no problem that can't be solved! In this division problem we will add 29 to itself as many times as we need to get to 124. Remember, multiplication is just adding but written differently!

Now, we can stop here because we know that 116 + 29 will equal more than 124. So we put the 4 on top and subtract the 116 from the target of 124 to find the remainder.

The answer to 124 ÷ 29 is 4r8

**NICELY DONE!**

I believe you understand what we just did!

Remember that division is a multiplication problem written differently. Multiplication is an addition problem written differently.

**REVIEW: Division Examples**

| YOU SEE | I SEE | I REWRITE |
|---|---|---|
| 12 ÷ 4 = | 4 x what number = 12 | 4 $\overline{)12}$ |

It's highly important to remember the following:

If you need to know what the HALF of any number is, divide it by 2!

So to find half of 10, divide 10 by 2. To find half of 80, divide 80 by 2.

EXAMPLE:

$$\begin{array}{ccc} & 5 & & 40 \\ \text{Half of 10} & 2\overline{)10} & \text{Half of 80} & 2\overline{)80} \end{array}$$

# 10

# FRACTIONS MADE SIMPLE

Okay, here we go...

Fractions are something that can really freak people out. I remember fractions as a scary thing. However, I used to kinda cheat my way through these, yet I was still doing all the math. Just differently!

KEEP IN MIND there are two different ways to write a fraction: You can stack them using a numerator and a denominator. Then, there is the DECIMAL way of showing a fraction. I used to turn all my stacked fractions into decimal fractions, then do the calculations (the math), and at the end, I'd change them all back to the "stacked" format. I just always found it easier to do it that way! You might like doing it the other way... "to each his own!

I'm going to show you DECIMAL fractions and then stacked fractions (numerator and denominator fractions).

Before we get into the actual math of things, it is important that you know the following:

**When there is a decimal point shown, the amount of zeros after the decimal point can go on and on and on.**

## DECIMAL POINT
### 0005.000

This number is actually read as five (5). The zeros to the left of the five have no value. They are just space fillers. This is why we don't write them. The same thing goes for the zeros to the right of the decimal. They are being used as space fillers. They have no value.

The numbers to the right of the decimal will be used when adding, subtracting, and multiplying, or dividing  with mixed numbers  or numbers with fractions. You'll catch on in just a bit, I promise!

|      | 100s | 10s | 1s | Decimals | 10ths | 100ths | 1,000ths |
|------|------|-----|----|----------|-------|--------|----------|
| EX 1 |      |     | 9  | .        | 1     |        |          |
| EX 2 |      | 1   | 5  | .        | 1     | 2      | 5        |
| EX 3 |      | 1   | 0  | .        | 2     | 5      |          |

**The numbers above usually look the following:**

**9.1**       9.1 is read...       "9 point 1" or "9 and $^1/_{10}$"
**15.125**    15.125 is read...    "15 point 125" or "15 and $^{125}/_{1000}$"
**10.25**     10.25 is read...     "10 point 25" or "10 and $^{25}/_{100}$"

You may be asking, "Why is this important?"

When solving math problems, it's important to know what's to the left and what's to the right of the decimal, so we can come up with the absolute correct answer.

**EXAMPLE:**
10.5       is the same as:  $10\,^5/_{10}$     reduced:  $10\,^1/_2$
9.25       is the same as:  $9\,^{25}/_{100}$     reduced:  $9\,^1/_4$
6.125      is the same as:  $6\,^{125}/_{1000}$     reduced:  $6\,^1/_8$

I try to do all my figuring (math solving) by using decimal fractions. I don't like doing anything with stacked fractions (numerator and denominator). They just don't make sense to me.

When adding numbers with decimals, it is important to keep your numbers lined up straight.

Example 1    Example 2

| 10.5 | 20.3 | rewrite this problem: | 20.30 | By adding a zero. you didn't |
|---|---|---|---|---|
| +15.2 | +25.35 | | +25.35 | change the number. |
| 25.7 | | | 45.65 | **ANSWER** |

Notice in Example 2, we added a zero to 20.3 to make it read 20.30. We did this to keep the numbers straight...to keep the decimal point lined up neatly. Then we just added the numbers. Remember, a zero has NO VALUE.

Okay, this works the same for subtracting numbers with decimal points as well. Let's take a look:

10.5
−10.25

As we can see, in the example above, the decimals did not line up. So, what did we have to do? Correct, rewrite the equation so the decimals do line up...then we solve! Let's look at another example.

| 10.50 | 125.35 | Looks easy enough to solve, right? Not So Fast! We need to rewrite the math problem so the decimal points line up. I think we need to add a 0 to the right of 10.5. This makes the number 10.50...now the decimals line up. We're ready to solve! |
|---|---|---|
| −10.25 | −5.6 | |

| 125.35 | This looks easy enough to do the math problem, but we still need to rewrite so the decimal points line up! I think we should Add a 0 next to the 6. |
|---|---|
| −5.60 | |

**Remember, we are not ready to solve an addition or subtraction problem until the decimal points are lined up!**

You may be thinking to yourself, *When will I ever use this!?* Well, if I told you that you use it almost every day, would you believe me? Let's look at what I am talking about!

Quick question: Have you ever been walking around and found a penny ($0.01 or 1¢), quarter ($0.25 or 25¢), dime ($0.10 or 10¢), or nickel ($0.05 or 5¢)? Do you ever pick it up and put it in your pocket? If so, this is where the decimal comes in every day! Every time you have **American Money**, you use decimals!

When you hold up a dollar bill and look at it, you see one dollar. When you write it down, it usually looks like this: **$1.00.**

Let's take a look at what we just wrote!

First, "$" this symbol stands for "DOLLARS," or another way to say it— MONEY.

$1.00

"1" stands for how many dollars we have.
Next is the DECIMAL.
Then, the place for change to be written.

When writing or looking at money, the paper $1.00 bill has more value than the coins:

Which is worth more, a paper dollar bill or 10 pennies?
**$1.00 or 10¢ ($0.10)**?   A dollar bill.
Which is worth more, a paper dollar bill or one nickel?
**$1.00 or 5¢ ($0.05)**?   A dollar bill.
Which is worth more, a paper dollar bill or one dime?
**$1.00 or 10¢ ($0.10)**?   A dollar bill.
Which is worth more, a paper dollar bill or one quarter?
**$1.00 or 25¢ ($0.25)**?   A dollar bill.

So, keeping with money, the penny, nickel, dime, and quarter are essential. These coins are pieces of a dollar. The coins are commonly called "change." But in reality, each **coin represents a fraction of a dollar—in other words, a percentage of a dollar.**

Let's continue looking at the dollar.

|  | 1s | 10ths | 100ths |
|---|---|---|---|
| $1.53 | 1 | 5 | 3 |

1s = 1 whole dollar
10ths = dimes
100ths = pennies

**Breaking $1.53 Down:**

**$1 = One Whole dollar.** Another way to say it is 100 pennies!

**The Decimal separates the whole dollars from the change.**

**.53 = 53 cents** = 53 pennies, or 5 dimes and 3 pennies, or 2 quarters and 3 pennies.

Remember this chart from earlier? Let's put some money on this

|  | 1,000s | 100s | 10s | 1s | Decimal | 10ths | 100ths | 1,000ths |
|---|---|---|---|---|---|---|---|---|
| $ |  |  |  |  | . |  |  |  |
| $ |  |  |  |  | . |  |  |  |
| $ |  |  |  |  | . |  |  |  |
| $ |  |  |  |  | . |  |  |  |

Read each amount (number) below and write it on the chart. We'll do the first one!

$1,859.84
$512.37
$14.50
$50.25

|  | 1,000s | 100s | 10s | 1s | Decimal | 10ths | 100ths | 1,000ths |
|---|---|---|---|---|---|---|---|---|
| $ | 1 | 8 | 5 | 9 | . | 8 | 4 |  |
| $ |  |  |  |  | . |  |  |  |
| $ |  |  |  |  | . |  |  |  |
| $ |  |  |  |  | . |  |  |  |

Now we know we will **not need** 1,000ths place when dealing with money until much later in life. So, if you'd like, you can put a zero in the 1,000ths column, When doing math problems, you may be asked to use the 1000ths column.

Now, when adding money, you don't want to make a mistake. So you have to have all your numbers in a row. You have to keep your columns straight. Even if you're not adding money and you're just adding numbers, you have to keep your numbers straight and lined up properly...especially when there is a decimal. Your decimals have to be lined up!

**EXAMPLE:**
**55.5 + 1 + 6.255 + 10.725 + 100**
This looks kinda hard to solve. A little tricky, right?
 Well, it's just an addition problem. Remember, every problem has a solution!
**Let's work this one together!**

**Step one:** Rewrite the math problem into a stacked form. Keep your numbers straight!

| | |
|---|---|
| 55.5 | Now looking at this addition problem as it is written now, it looks a little easier. But, here's the thing, the way it is written we'll be trying to add whole numbers to |
| +1 | fractions...you cannot do this, the answer <u>will not</u> be correct! So, what we have to do, is count how many places to the right of the decimal is used and make all the |
| +6.255 | numbers go out to there by adding ZEROs. We have to look for the number with the most numbers to the right of the decimal which is **10.725**. Now we know that |
| +10.725 | all the numbers in the addition problem should go out three places, to the **1,000ths place.** |
| +100 | That's right, we have to rewrite this darn problem to make sense! |

| | |
|---|---|
| 55.500 | |
| +1.000 | Now look at the math problem...a lot simpler, right? Now the first thing we do is to |
| +6.255 | know that the answer will go out to the 1,000ths. Bring the decimal point straight down! |
| +10.725 | |
| +100.000 | |

**Like I said early on, I prefer the largest number on top. So, I rewrite again!**

| | |
|---|---|
| 100.000 | |
| +10.725 | |
| +55.500 | |
| +1.000 | |
| +6.255 | **Always remember where the decimal goes! We figured this problem to the 1,000ths!** |
| 173.480 | |

**Include the zero 0 in the answer to show that you took the answer to the 1,000ths! Also, it helps you to keep the answer straight.**

Now, let's use money to show why it's necessary to line up your decimals. Why the decimal point is so important! Let's add the following moneys:

$1.25        $2.50        $3        $2.25        $2

34

First, rewrite the math (money) problem.

**If your numbers aren't lined up properly and you add, you'll come up with the wrong answer!**

| | |
|---|---|
| $1.25 | **So,** first you have to rewrite the money problem to include all the amounts to the 100ths. **Now**, when adding these numbers as written, the decimals are not in line. In fact, there is **no decimal shown for the $3 and the $2!** Remember, the decimal separates the whole **dollar** from the change. So, if you have enough change, it makes a whole dollar! Now add! |
| +$2.50 | |
| +$ 3 | |
| +$2.25 | |
| +$2.00 | |
| $6.05 | |

**Now, we'll redo this and add decimals for the $3 and $2 and see what we come up with!**

This time, as you read the numbers, include the cents—$3 and zero cents; $2 and zero cents. Write down what you read.

| | |
|---|---|
| $1.25 | 1 dollar and 25 cents. |
| +$2.50 | 2 dollars and 50 cents. |
| **+$3.00** | **3 dollars and 00 cents.** |
| +$2.25 | 2 dollars and 25 cents. |
| +$2.00 | **2 dollars and 00 cents.** |
| $ 11.00 | |

**Now,** as you can see, written properly, the sums are very different. This is why it is so important to line up your numbers properly and always have the decimal points lined up properly!

**I don't know about you, but I would much rather have $11.00 than $6.05! Knowing is half the battle!**

The same thing goes for subtracting numbers with decimals. You have to keep the decimals in line.

EXAMPLES:

1) 12.25 – 10.5 =
2) 15.80 – 12.8 =
3) 10 – 3.5 =
4) 12.65 – 2.5 =

So, let's set these numbers up for subtraction!

Read the first subtraction problem out loud as you write down the numbers.

1) "12 point 25 minus 10 point 5-0 equals..."

| | WHOLE NUMBERS | PARTS (FRACTIONS) | |
|---|---|---|---|
| 12.25 | 12. | 25 | We know that we cannot subtract 50 from 25, so, we have to take 1 from the 12 and put it next to the 25, making it 1.25. |
| −10.50 | −10. | −50 | |
| | WHOLE NUMBER | PARTS (FRACTIONS) | |
| | 11. | 1.25 | |
| | −10. | −50 | Squeeze the answer together: |
| | 1. | 75 | 1.75 ! |

2) "15 point 8 minus 12 point 8 equals…"

|        | WHOLE NUMBERS | PARTS (FRACTION) |
|--------|---------------|------------------|
| 15.8   | 15.           | 8                |
| −12.8  | −12.          | 8                |
| 3.0    | 3.            | 0                |

3) "10 minus 3 point 5 equals…"

|        | WHOLE NUMBERS | PARTS (FRACTIONS) |                          |
|--------|---------------|-------------------|--------------------------|
| 10.0   | 10.           | 0                 | We know we have to rewrite the problem. |
| −3.5   | −3.           | 5                 |                          |
|        | WHOLE NUMBERS | PARTS (FRACTIONS) |                          |
|        | 9.            | 1.0               |                          |
|        | −3            | −0.5              | Now for the good part: squeeze the numbers together! |
|        | 6.            | 5                 |                          |

**6.5** is the answer. **GOOD JOB!**

During these subtraction problems, we found the solutions by breaking the numbers into two separate problems. We subtracted the whole numbers, and we subtracted the "parts" (fractions). Once we found the solutions to both, **WE SQUEEZED THE NUMBERS TOGETHER!**

4) "12 point 6-5 minus 2 point 5 equals..."

|  | WHOLE NUMBERS | PARTS (FRACTIONS) |  |
|---|---|---|---|
| 12.65 | 12. | 65 |  |
| −2.50 | − 2. | −50 | Now, what do we have to do? That's right! Squeeze the |
|  | 10. | 15 | numbers together! |

**10.15!**

**Nicely Done!**

# 11

# MORE ON DECIMALS

In life, as well as mathematics, you'll face many problems, many challenges. RE-MEMBER, for every problem, every challenge, there is a SOLUTION! If you don't get it right the first time, you have an eraser...use it!

**Now, we're switching from addition and subtraction problems to Multiplication Problems!**

**Multiplying Decimals:**

6.5 x 2.5 =

The first thing we do is to stack the multiplication problem.

| | |
|---|---|
| 6.5<br>x 2.5 | With the case of fractions (decimals), we have to multiply it out. We will work this one out together. |

# TRADITIONAL METHOD:

|  |  | 6.5 |  |  |
| --- | --- | --- | --- | --- |
|  |  | x 2.5 |  |  |
| **Step One:** | 5x5 | 5 | Carry the 2 |  |
| **Step Two:** | 5x6+2 | 325 |  | Steps 1 and 3 |
| **Step Three:** | 2x5 | 0 | Carry the 1 |  |
| **Step Four:** | 2x6+1 | +130 |  |  |
| **Step Five:** | Add Step Two to Step Four:   325                                              +130 | 1625 |  |  |
| **Step Six:** | This might confuse you...**PLACING THE DECIMAL POINT!** To get this right, you have to count how many numbers are to the right of all numbers in the multiplication problem. This will tell you how many spaces you will have to move the decimal place to the left! |  |  |  |

## Placing the Decimal Point Using the Example: 6.5 x 2.5

How many places to the left do we have to move the decimal? WELL, when we multiply 65 x 25, we get 1625. However, the problem is 6.5 x 2.5, so we have to place the decimal. We know that 6.5 has one number to the right of the decimal and so does 2.5. Therefore, we know that we have to move the decimal two places to the left. The answer to 6.5 x 2.5 is 1625, **and we place the decimal two places to the left: 16.25.**

## My Method

My method of doing this math problem is to break 6.5 x 2.5 into two smaller math problems and them add them together!

| 6.5 | and | 6.5 | |
|---|---|---|---|
| x 2 | | x 0.5 | |
| 13.0 | and | 3.25 | Now, add these two numbers to get the answer! |

| 13.00 |
|---|
| + 3.25 |
| 16.25 |

Let's do another one:

25.25 x 4 =

| 25 | .25 | |
|---|---|---|
| x 4 | x 4 | |
| 100 | 1.00 | Now add the answers together! |
| +1 | | |
| **101** | | |

You may be scratching your head trying to figure out what happened to the 101.**00** in the answer. Let me explain! The ".00" has no value at all. You don't have to show it unless you are asked to show your answer to the nearest hundredths.

Or if it makes you feel better to add the **.00** to your answer, by all means put it in!

**So you fully understand about value, let me explain** a little more. Let's take a look at the numbers below. In the examples below, the equal sign (=) means "the same as":

1.500000 = 1.5          12.00000 = 12     5.00 = 5.0 = 5
0001.00 = 1     05.5000 = 5.5       050.00100 = 50.001      20.100 = 20.1

I hope this makes it clear about the zeros at the beginning or end of a number.

**Dividing Decimals:**

In these examples, we will not use remainders. **We will use decimals!**

1) $3 \div 6 =$
2) $4 \div 12 =$
3) $6 \div 12 =$

1) We read the first problem: "3 divided by 6."
   Let's get to it!

| | |
|---|---|
| $6\overline{)3}$ | As we can see, 6 does not go into 3. So, we know the answer will be a fraction! |
| $6\overline{)3.0}$ | We can add a decimal with a 0 following it because the zero has no value. The number is still "3" However, when we add the decimal point and zero we can now divide. Please remember to put a decimal point on the answer bar just above where you added the decimal below. |
| $\begin{array}{r} 0. \\ 6\overline{)3.0} \end{array}$ | So, to keep it straight, place a 0. over the 3 because we know that 6 goes into 3, 0 times. |
| $\begin{array}{r} 0.5 \\ 6\overline{)3.0} \end{array}$ | Now, just divide like normal. 6 goes into 30 five times.  Write it down...and that's your answer: 0.5! |

2) We read: "4 divided by 12."

| | |
|---|---|
| $12\overline{)4}$ | |
| $\begin{array}{r}0.\phantom{0}\\12\overline{)4.0}\end{array}$ | Just by looking at this math problem we know we need to add a decimal and a 0. Remember to bring the decimal up! |
| | |
| $\begin{array}{r}0.33\phantom{0}\\12\overline{)4.00}\end{array}$ | The next step is to add another 0 to the 4.0 and drop it down to the next 4. |
| $\underline{-36}$ | |
| $40$ | |

As you can see, 12 goes into 4, .3333 times. 12 into 40 will always be 3 with 4 left over. Therefore, take your answer to the 100ths and round off...THAT'S IT! Good going! I trust you are feeling a bit more confident that you know about decimals and how they are used.

3) We read: "6 divided by 12."
Do the same thing we did in the past two examples!

| | |
|---|---|
| STEP ONE: | $12\overline{)6}$ |
| STEP TWO: | $\begin{array}{r}0.\phantom{0}\\12\overline{)6.0}\end{array}$ |
| STEP THREE: | $\begin{array}{r}0.5\phantom{0}\\12\overline{)6.0}\end{array}$ |
| | $\underline{-60}$ |
| | $0$ |

Now, we're going to tackle a more difficult concept.

**Here's our next math problem:**

**8.5 ÷ 4.25 =**

As you **can** see, both numbers have the decimal.

**WE CANNOT DIVIDE A NUMBER BY A FRACTION. JUST KNOW, IF YOU MULTIPLY ONE NUMBER BY 10, THEN YOU HAVE TO MULTIPLY THE OTHER NUMBER BY 10!**

| | |
|---|---|
| **Remember,** you cannot divide a number by a number with a decimal in it. | $4.25\overline{)8.5}$ |
| We have to multiply each number by 10 to move the decimal one place to the right. | $42.5\overline{)85}$ |
| Then we have to do it again until the second decimal is gone in the outside number! | $425\overline{)850}^{\,2}$ |
| The Answer to 8.5÷4.25 is | **2!** |

SEE HOW THE OUTSIDE NUMBER HAS NO MORE DECIMAL? NOW YOU CAN DO THE MATH!

REMEMBER THIS FROM EARLIER?

| 1,000s | 100s | 10s | 1s | DECIMAL | 10ths | 100ths | 1000ths |
|---|---|---|---|---|---|---|---|
| | | | | . | 1 | 2 | 5 |

**You're going to learn how to turn a decimal fraction into a numerator/denominator fraction.**

As you can see, the number above is .125. That number is 125 thousandths. To make it into the other type of fraction, simply do the following:

| $\frac{125}{1000}$ | now reduce | $\frac{5}{40}$ | reduce again | $\frac{1}{8}$ |
|---|---|---|---|---|

| 1,000s | 100s | 10s | 1s | DECIMAL | 10ths | 100ths | 1000ths |
|---|---|---|---|---|---|---|---|
| | | | | . | 2 | 5 | |

As you can see, the number above is .25. The number is read as 25 hundredths. To make it into the other type of fraction, simply do the following:

| $\frac{25}{100}$ | now reduce | $\frac{1}{4}$ |
|---|---|---|

| 1,000s | 100s | 10s | 1s | DECIMAL | 10ths | 100ths | 1000ths |
|---|---|---|---|---|---|---|---|
| | | | | . | 4 | | |

As you can see, the number above is .4. The number is read as 4 tenths. To make it into the other type of fraction, simply do the following:

| $\frac{4}{10}$ | now reduce | $\frac{2}{5}$ |
|---|---|---|

We haven't gone into what REDUCE means. However, you have all the tools to REDUCE a fraction!

TO REDUCE means to make the numerator and denominator the smallest it can be. The only way to do this is to KNOW that whatever you do to the top number, you have to do to the bottom number!

## EXAMPLE 1:

| $\frac{18}{36}$ | Since both numbers are even numbers, you know you can reduce by 2. | $\frac{18 \div 2 =}{36 \div 2 =}$ | $\frac{9}{18}$ |
|---|---|---|---|
| $\frac{9}{18}$ | Reduce again by 3. | $\frac{9 \div 3 =}{18 \div 3 =}$ | $\frac{3}{6}$ |
| $\frac{3}{6}$ | Reduce again by 3. | $\frac{3 \div 3 =}{6 \div 3 =}$ | $\frac{1}{2}$ |
| $\frac{1}{2}$ | | | |

## EXAMPLE 2:

| $\frac{15}{20}$ | REMEMBER the RULE OF 5!<br>If a number ends with a 0 or a 5, it can be evenly divided by 5! |
|---|---|
| $\frac{3}{4}$ | |

## EXAMPLE 3:

Here's a hint and a rule. If all the digits of a number add up to a number that is divisible by three, then the number is divisible by 3!

| | |
|---|---|
| $\dfrac{42}{96}$ | Like 42. 4+2=6 And 6÷3=2. Therefore 42 is divisible by 3! 42÷3=14. Like 96. 9+6=15 and 15÷3=5. Therefore 96 is divisible by 3! 96÷3=32. |
| | Reduce by 3. |
| $\dfrac{14}{32}$ | Both 14 and 32 are even numbers. Therefore, they are divisible by 2. Reduce the fraction by 2. |
| $\dfrac{7}{16}$ | 7/16 is the answer! |

**Keep in mind that fractions are just division problems written differently.**

| | | | |
|---|---|---|---|
| $6 \div 2 =$ | is the same as... | $\dfrac{6}{2}$ | Both are solved the same way. |
| | | $2\overline{)6}^{\,3}$ | |

| | | |
|---|---|---|
| $\dfrac{1}{2}$ | is the same as... | $2\overline{)1}$ |

We know that 2 cannot go into 1. However, looking at this problem, we know we can add a decimal and a zero.

| | | | | | |
|---|---|---|---|---|---|
| Remember to move the decimal point up: | $2\overline{)1.0}^{\,0.50}$ | 0.5 = | $\dfrac{5}{10}$ | REDUCE | $\dfrac{1}{2}$ |

When adding or subtracting fractions you have to see the numbers and realize what's happening with the fractions. (Also, when adding and subtracting fractions, the denominators have to be the same!)

Let's take a look at what I mean:

| $\frac{1}{2}$ | + | $\frac{1}{2}$ | The denominators are the same, so we add the numerators. |
|---|---|---|---|
| | $\frac{2}{2}$ | | $\frac{2}{2}$ **This reduces to 1!** |

# 12

# PERCENTAGES MADE SIMPLE

The easiest way to talk about fractions (percent) and decimals is to talk about MONEY! Each dollar is 100 percent of 1 dollar. The change is just a fraction (percent) of a dollar! However, if you have enough change, it will total A DOLLAR ($1.00).

Let's look at some examples:

| 100 pennies = $1.00 | 20 Nickels = $1.00 | 4 quarters = $1.00 | 2 half-dollars = $1.00 |

A dollar is usually written out as $1.00. Look at the number $1.00. Remove the dollar sign '$' and what is left over? 1.00. How many pennies does it take to make a dollar? 100 pennies makes a dollar.

Remember the chart below from earlier? Let's put $1 on to this chart!

|     | 1,000s | 100s | 10s | 1s | DECIMAL | 10ths | 100ths |
|-----|--------|------|-----|-----|---------|-------|--------|
| $   |        |      |     | 1  | •       | 0     | 0      |

You can see how money is just a decimal number.

| 1 dollar = $1.00 | 1 half-dollar = $0.50 | 1 quarter = $0.25 |
| 1 dime = $0.10 | 1 nickel = $0.05 | 1 penny = $0.01 |

A quarter is called a quarter because 4 quarters (of anything) = 1 whole.

And, 4 quarters (the coin) = $1.00 .

| $4\overline{)100}^{\phantom{0}25}$ | $\dfrac{25}{100}$ | REDUCE | $\dfrac{1}{4}$ |
| --- | --- | --- | --- |

A fifty-cent piece is also known as a half-dollar. It takes two halves to make a whole. A half-dollar is written $0.50.

| $\dfrac{50}{100}$ | **REDUCE** | $\dfrac{1}{2}$ |
| --- | --- | --- |

NOW WE'VE JUST GOTTEN STARTED TALKING ABOUT MONEY. MONEY BRINGS US TO **PERCENT!**

Whenever I used to see the "**%**" symbol, I would say "UH–OH," and all I could see was my grade going down. Do you feel the same way? Well, now we're going to make sense about this whole thing!

All the % symbol means is "divided by 100." As explained above...**1 penny is also called 1 CENT! Another way of saying this is 1 per cent of a dollar!** Another way of writing it is $0.01.

When figuring out percentages, it is very important that you don't freak out! Just remember, you already know this stuff. You just don't know you know it... but you do!

Here we go!

PENNY: 1 cent; 1¢; ¢1; $0.01. A penny is 1 percent of a dollar. It takes 100 pennies to make a dollar.

A penny is 1 one-hundreth of a 1 whole dollar!

On the familiar chart below, you can see how 1 penny goes on the chart. Just squeeze the numbers together!

|  | 1,000s | 100s | 10s | 1s | DECIMAL | 10ths | 100ths |
|---|---|---|---|---|---|---|---|
| $ |  |  |  | 0 | • | 0 | 1 |

The information in the chart squeezed together is written **$0.01!** See how the 1 is in the hundredths place?

Now we're going back to the chart. Yes, the one that we keep using for decimals. However, we are adding a little more information to it. And, yes, we're still on money!

|  |  |  |  | 100% |  | 10% | 1% |
|---|---|---|---|---|---|---|---|
|  | 1,000s | 100s | 10s | 1s | DECIMAL | 10ths | 100ths |
| $1.00 |  |  |  | 1 | • | 0 | 0 |
| $0.65 |  |  |  | 0 | • | 6 | 5 |
| $0.50 |  |  |  | 0 | • | 5 | 0 |
| $0.15 |  |  |  | 0 | • | 1 | 5 |
| $0.10 |  |  |  | 0 | • | 1 | 0 |
| $0.05 |  |  |  | 0 | • | 0 | 5 |
| $0.01 |  |  |  | 0 | • | 0 | 1 |

Again, when we're discussing percentages, we write the percentages as follows:

| NUMBER | % of 1 | Decimal | Fraction |
|--------|--------|---------|----------|
| 1.00 | 100% | 1.00 | $\frac{100}{100}$ |
| 0.65 | 65% | .65 | $\frac{65}{100}$ |
| 0.50 | 50% | .50 | $\frac{50}{100}$ |
| 0.15 | 15% | .15 | $\frac{15}{100}$ |
| 0.10 | 10% | .10 | $\frac{10}{100}$ |
| 0.05 | 5% | .05 | $\frac{5}{100}$ |
| 0.01 | 1% | .01 | $\frac{1}{100}$ |

| | 10,000% | 1,000% | 100% | | 10% | 1% |
|--------|---------|--------|------|---------|------|------|
| | 100s | 10s | 1s | DECIMAL | 10ths | 100ths |
| 100% | | | 1 | ● | 0 | 0 |
| 1,000% | | 1 | 0 | ● | 0 | 0 |

| | 1,000% | 100% | | 10% | 1% | FRACTION |
|---|---|---|---|---|---|---|
| | 10s | 1s | DECIMAL | 10ths | 100ths | |
| 1000% | 1 | 0 | • | 0 | 0 | $\frac{1000}{100}$ |
| 100% | | 1 | • | 0 | 0 | $\frac{100}{100}$ |
| 65% | | | • | 6 | 5 | $\frac{65}{100}$ |
| 50% | | | • | 5 | 0 | $\frac{50}{100}$ |
| 25% | | | • | 2 | 5 | $\frac{25}{100}$ |
| 10% | | | • | 1 | 0 | $\frac{10}{100}$ |
| 5% | | | • | 0 | 5 | $\frac{5}{100}$ |
| 1% | | | • | 0 | 1 | $\frac{1}{100}$ |

When you see a %, don't freak out. It just means the following:

It's always over 100 when you see a %. So when you see 27%, think of it as $^{27}/_{100}$ (read as "27 over 100"). Or as a decimal fraction - 0.27.

| Percent | Fraction | REDUCE | Decimal |
|---------|----------|--------|---------|
| 1,000% | $\dfrac{1,000}{100}$ | $100\overline{)1,000}$ | 10.00 |
| 100% | $\dfrac{100}{100}$ | $100\overline{)100}$ | 1.00 |
| 50% | $\dfrac{50}{100}$ | $100\overline{)50.0}$ | 0.50 |
| 25% | $\dfrac{25}{100}$ | $100\overline{)25.00}$ | 0.25 |

You may be thinking to yourself , *Why does this matter? Who cares about percentages?*

Well, to answer these questions, you have to think about making a purchase—ya know, shopping! When you go to the Dollar Tree and you have a dollar with you.  You see a bag of candy you really want. You go to the cash register to pay for the candy. The cashier slides the bag of candies across the register.

The cashier says, "That'll be $1.05."

You only have $1.00 because you went to the Dollar Tree where everything is a dollar! Yes, the candies cost $1.00, but there's a 5% sales tax that you have to pay.

5% of a dollar is 5¢—a nickel, or 5 pennies. Tax rates are set by the government. The seller charges the 5% to the buyer, so they can make the sale for a dollar.

Okay, using what we just discussed, we'll work some percentage problems together!

*Sales tax is 8%. You want to buy something that costs $20.00. How much will you actually have to pay at the register?*

There are many ways to solve this problem! Let's stop talking about it and take a look!

Ultimately, we want to figure out what 8% of 20 is and then add it to 20. Then we will know what the true price is:

**$20.00 plus 8% for sales tax =**

| | | | | | | | |
|---|---|---|---|---|---|---|---|
| STEP 1: | 20 | | | | | | |
| STEP 2: | 8% | | | | | | |
| STEP 3: | $\frac{8}{100}$ | | | | | | |
| STEP 4: | $\frac{20}{1}$ | | | | | | |
| STEP 5: | $\frac{20}{1}$ | x | $\frac{8}{100}$ | = | $\frac{160}{100}$ | = | $100\overline{)160.00}$   1.60 |
| STEP 6: | Add $20.00 + $1.60 = $21.60 | | | | | | |

You'll need $21.60 to purchase the $20.00 item because of taxes.

The way we just did the problem above, I always found that I would easily get lost, so I decided to go another way.

**My Method** of doing it, I call "GO WITH WHAT YOU KNOW"

Let me explain this method of figuring percentages! When you multiply any number by 10, all you have to do is to move the decimal one place to the right:

50 x 10 = 500
13 x 10 = 130
5 x 10 = 50.

You understand this, right?
So, it only makes sense that when you multiply any number by 'one-tenth' or 10% or 0.1, all you have to do is move the decimal one place to the left:

50 x 0.10 = 5.0
13 x 0.10 = 1.3
5 x .10 = 0.5!

It also makes sense that if you know what 10% of a number is, you know what 1% of a number is. To find 1% of a number, all you have to do is move the decimal two places to the left:

50 x .01 = 0.5
13 x .01 = 0.13
5 x .01 = .05

Get it? Just move the decimal two places to the left!

Congratulations! You just learned the fastest way to do percentages!

Let me explain using the same problem as before:

*$20 item plus 8% sales tax. How much money do you need to buy it?*

> Step One: You already know that 1% of 20 is 0.2
> Step Two: So 0.2 x 8 = 1.60
> Step Three: 20 + 1.60 = 21.60
> Step Four: **Label your answer. $21.60.**

When looking for 25% of a number, I know that 25% = $^1/_4$ = 0.25. I know that to find 25% of any number, I just have to divide that number by 4! Think of money. How many quarters does it take to make $1.00? It takes 4. A quarter is written like this 0.25 or 25%.

YEP! You got it. Another chart!

| Number | 50% | 25% | 10% | 1% | .5 or half . | 25 or quarter |
|---:|---:|---:|---:|---:|---:|---:|
| 100 | 50 | 25 | 10 | 1 | 50 | 25 |
| 40 | 20 | 10 | 4 | 0.4 | 20 | 10 |
| 20 | 10 | 5 | 2 | 0.2 | 10 | 5 |
| 10 | 5 | 2.5 | 1 | 0.1 | 5 | 2.5 |

**EXAMPLES OF "GOING WITH WHAT YOU KNOW!"**

1) What is 14% of 50?

I know that 10% of 50 is 5. I also know that 1% of 50 is 0.5. You may be asking yourself, *So what?*

Well... 14 = 10 + 4.

So let's get to it:

| | | |
|---|---|---|
| 10% | = | 5.0 |
| 1% | = | + 0.5 |
| 1% | = | + 0.5 |
| 1% | = | + 0.5 |
| 1% | = | + 0.5 |
| | | **7.00** |

**This is all about how to make a difficult math problem EASY!**

2) What is 32% of 145?

The way I would do this math problem is easy for me. I already know that 10% of 145 is 14.5. I know that 1% of 145 is 1.45. The question is, what is 32% of 145?

I know that 32 is three tens and two ones. So I add the three tens and the two ones, then add those numbers together to get 32%.

| | | | |
|---|---|---|---|
| 14.5 | 1.45 | 43.5 | |
| +14.5 | + 1.45 | + 2.9 | By going with what I know and doing simple addition, I |
| +14.5 | 2.90 | 46.4 | know that 32% of 145 = 46.4 |
| 43.5 | | | |

KEEP IN MIND if the math problem is simple enough for you, you have practiced what you are learning, and you can do the math in your head, GO FOR IT!

We're gonna change up the question a little...

3) 100 x 15% =

This is the same as being asked, "What is 15% of 100?"
When you see a question like this, GO WITH WHAT YOU KNOW!

| 10% of 100 | 5% of 100 | |
|---|---|---|
| 10 | 1 | |
| | +1 | Like I said a moment ago, if the math is easy enough, you can do the math in your head. We knew that 1% of 100 is 1. We knew we needed 5 ones, so we add 1% five times. We also knew that we needed 10% of 100, and we got that. Now we just add the 10% and 5% together to get the answer! |
| | +1 | |
| | +1 | |
| | +1 | |
| | 5 | We add both together, and we get 15. So, **we know that 15 is 15% of 100.** |

4) 80 − 20% =

First thing we need to do is KNOW that we have to find 20% of 80. Then we have to subtract that number from 80 to find the answer. So, going with what you know is the best way to solve this. We know that 10% of 80 is 8. So to know what 20% is, we ADD 8+8, which equals 16. Then we have to subtract 16 from 80 and that will be the answer.

| |
|---|
| 8 |
| +8 |
| 16 |
| |
| 80 |
| −16 |
| 64 |

Now here comes a real DOOZIE!

5) What is 20% of 100 − 80% of 20?

What do you think we need to do first? Think about it.

Okay, now that you have given it some thought, the first thing we need to do is to find out what 20% of 100 is. The second thing we need to do is to find out what 80% of 20 is. Let's get started!

We know that 10% of 100 is 10. Therefore 10 + 10 = 20. Now we know that 20% of 100 is 20. Write it down.

20

Now we need to find 80% of 20.

Well, we know that 10% of 20 is 2. We can add 2 eight times, or multiply 2 times 8.

Either way 80% of 20 = 16! Write it down.

```
   20
  −16
    4
```

4 is the answer to the difficult question!

Since that one went so well, we're gonna do another one! Remember, GO WITH WHAT YOU KNOW!

## 6) What is 15.5% of 30?

| 10% of 30 | 1% of 30 | Here's the tricky part. Now we have to figure out what .5% of 30 is. You already know that 1% of 30 is .3, so to find .5% of 30, divide .3 by 2, and you get .15. |
|---|---|---|
| 3 | 0.3 | |
| | | $\begin{array}{r} .15 \\ 2\overline{)0.30} \end{array}$ So we know that .5% of 30 is 0.15. We also know that 10% of 30 is 3, and that 5% of 30 is half of 3, or 1.5. |

Now is all we have to do is add the numbers together to get the answer. Let's do it!

| |
|---|
| 3.00 |
| +1.50 |
| +0.15 |
| 4.65 |

Now we know 15.5% of 30 is 4.65.

# IN CONCLUSION

# FROM ONE MATH STRUGGLER
# TO ANOTHER...

I have no more fast hints that I can give you at the present time. The rest, you will be able to pick up as you progress through your mathematics education. I created this system for ME, and it has been a pleasure sharing it with YOU! This is how I turned around my math life. If this helps you at all, then it was highly worthwhile for me to share.

The thing is, mathematics is difficult UNTIL YOU GET IT! Once you get it, it will be a breeze! As you develop your own shortcuts and different ways to look at a math problem, you'll be able to pass your hints and secrets to the next group of struggling students!

If you only need help with adding, it's in here. Same thing with subtraction, multiplication, and division. Some stuff, you'll have to learn at school. In this book, you'll find things from start-up math to more advanced math.

The hints in here will make ALGEBRA a piece of cake!

REMEMBER, begin your journey by nailing down the art of counting to 100 by 1s, 2s, 5s, and 10s! Then practice counting backwards from 100 to 0 by 1s, 2s, 5s, and 10s. The better you are at counting, the easier math will become!

The more practice you get, the quicker you will be able to visualize your numbers. Once you can visualize the numbers, the easier this all becomes!

| TRUST YOURSELF | | | | |
|---|---|---|---|---|
| GO WITH WHAT YOU KNOW | | | | |
| SUCCESS | SUCCESS | SUCCESS | SUCCESS | SUCCESS |